COMPLETE CHICKEN KEEPING JOURNAL

OWNER INFORMATION

NAME:

FARM NAME:

ADDRESS:

PHONE NUMBER:

LOG BOOK INFORMATION

Start Date:

End Date:

IF FOUND, PLEASE CALL OR

GOOD PRACTICES FOR CHICKEN FARM

LOCATION

- Ensure that farm is located away from any abattoir to prevent likely occurrence of diseases.
- Do not locate the farm in areas that are exposed to physical and chemical contamination that may be harmful to the animal health.

HOUSING

- House must be strong easy to clean and maintain.
- House floor must be concrete that is not too rough and must have adequate slope to ease cleaning.
- Remove every object that may lead to accident in the house.

FEEDING

- Feed must be of high quality, appropriate nutrition qualifying for nutritional requirement of age.
- Medicated feeds must be put separately from other feed with appropriate label.
- Provide separate place for animal feeding with good ventilation to prevent food contamination.

MANAGEMENT

- Chicken groups must have a specific worker attending to the needs to prevent misadministration of Vaccines.
- Farm worker must be well trained on farm management, chicken raising and feed management.

ANIMAL HEALTH

- Good hygienic practices must be implemented in farm to prevent accumulation of pathogens.
- chicken health must be regularly checked.
- Sick chickens must be separated from herd for medication to prevent spread of disease to others.
- Treatment of chicken must be under close supervision of licensed animal doctor.

ENVIRONMENT

- Always clean animal waste to prevent accumulation which can lead to offensive odors.
- Farm must have good sewage management system for improving quality of water waste.

HOW TO USE THE CHICKEN RECORD KEEPING LOG BOOK

This Chicken Log book is the first of its kind to offer in-depth approach and practical content on Chicken farm record keeping. It is specially designed for poultry owners and backyard Chicken keepers by an **experienced Chicken keeper** to record every activities and information of Chickens. In this book, you can record and analyze your observations on feeding, expenses, income, health, egg production, and other vital information for **4 years!** .

SECTIONS OF THIS LOG BOOK

Chicken Record : In this section you can record information of every chickens on your farm
You can track info such as name, ID, breed, appearance/color. purchase date, DOB, lay date, Cost, and notes

Income & Expenses section helps you track every info on regular expenditure and income of every month, you can record Date, item, unit/quantity, category, cost/price, profit, and others

Egg production : Track egg production daily, record daily egg count

Feed Tracking: on this page, you will record your feed combination strategy. Record the Period/Week, feed combination formula, brand, quantity and your observations.

Health Record | Medical Treatment: Record : Date, chicken ID, symptoms/injury/illness, medication, dosage, and other notes

Coop Maintenance : It's vital to keep your chicken coop clean, here, you can track, regular cleaning of your flock poultry. Record : Date, Deep/Light clean? and notes

Monthly Summary Sheet : Summarize your monthly activities and plan towards the next month on this page. Record your total monthly : Expenses, income, profit/loss, newly added chickens, mortality, and remark

Supplier and Veterinary Doctors Record : This is the section where you can record contact information of
important businesses, farm suppliers, veterinary doctors, and more.

Farm Detail : Record Farmer's name, farm name, address, phone number, and others

SUPPLIER CONTACT DETAILS

Name	
Business	
Phone	
Address	
Note	

Name	
Business	
Phone	
Address	
Note	

Name	
Business	
Phone	
Address	
Note	

Name	
Business	
Phone	
Address	
Note	

Name	
Business	
Phone	
Address	
Note	

Name	
Business	
Phone	
Address	
Note	

Name	
Business	
Phone	
Address	
Note	

Name	
Business	
Phone	
Address	
Note	

Name	
Business	
Phone	
Address	
Note	

Name	
Business	
Phone	
Address	
Note	

YEAR 1

MY FLOCK

Name	ID	Breed	Appearance	Purchase Date	DOB	Lay Date	Cost	Note

MONTH:

EXPENSES

Date	Item	Unit	Category	Cost
TOTAL				

INCOME

Date	Item	Unit	Category	Profit
TOTAL				

DATE	FURTHER NOTE | REMARK

Month :

EGG PRODUCTION

Day	No of chicken	Dark Brown	Light Brown	Blue	White	Green	NOTES
1							
2							
3							
4							
5							
6							
7							
8							
9							
10							
11							
12							
13							
14							
15							
16							
17							
18							
19							
20							
21							
22							
23							
24							
25							
26							
27							
28							
29							
30							
31							
Total							

FEED HISTORY | COMBINATION

Week/Period	Feed Combination	Brand Name	Quantity	Observation

HEALTH RECORD

Day	ID	Injury/Illness	Medication	Dosage	Note

COOP MAINTENANCE

Date	Deep Clean	Light Clean	Notes

DATE	FURTHER NOTE | REMARK

MONTH:

EXPENSES

INCOME

Date	Item	Unit	Category	Cost
TOTAL				

Date	Item	Unit	Category	Profit
TOTAL				

DATE	FURTHER NOTE | REMARK

Month :

EGG PRODUCTION

Day	No of chicken	Dark Brown	Light Brow n	Blue	White	Green	NOTES
1							
2							
3							
4							
5							
6							
7							
8							
9							
10							
11							
12							
13							
14							
15							
16							
17							
18							
19							
20							
21							
22							
23							
24							
25							
26							
27							
28							
29							
30							
31							
Total							

FEED HISTORY | COMBINATION

Week/Period	Feed Combination	Brand Name	Quantity	Observation

HEALTH RECORD

Day	ID	Injury/Illness	Medication	Dosage	Note

COOP MAINTENANCE

Date	Deep Clean	Light Clean	Notes

MONTH:

EXPENSES

Date	Item	Unit	Category	Cost
TOTAL				

INCOME

Date	Item	Unit	Category	Profit
TOTAL				

DATE	FURTHER NOTE | REMARK

Month :

EGG PRODUCTION

Day	No of chicken	Dark Brown	Light Brow n	Blue	White	Green	NOTES
1							
2							
3							
4							
5							
6							
7							
8							
9							
10							
11							
12							
13							
14							
15							
16							
17							
18							
19							
20							
21							
22							
23							
24							
25							
26							
27							
28							
29							
30							
31							
Total							

FEED HISTORY | COMBINATION

Week/Period	Feed Combination	Brand Name	Quantity	Observation

HEALTH RECORD

Day	ID	Injury/Illness	Medication	Dosage	Note

COOP MAINTENANCE

Date	Deep Clean	Light Clean	Notes

MONTH:

EXPENSES

INCOME

Date	Item	Unit	Category	Cost
TOTAL				

Date	Item	Unit	Category	Profit
TOTAL				

DATE	FURTHER NOTE | REMARK

Month :

EGG PRODUCTION

Day	No of chicken	Dark Brown	Light Brow n	Blue	White	Green	NOTES
1							
2							
3							
4							
5							
6							
7							
8							
9							
10							
11							
12							
13							
14							
15							
16							
17							
18							
19							
20							
21							
22							
23							
24							
25							
26							
27							
28							
29							
30							
31							
Total							

FEED HISTORY | COMBINATION

Week/Period	Feed Combination	Brand Name	Quantity	Observation

HEALTH RECORD

Day	ID	Injury/Illness	Medication	Dosage	Note

COOP MAINTENANCE

Date	Deep Clean	Light Clean	Notes

MONTH:

EXPENSES

INCOME

Date	Item	Unit	Category	Cost
TOTAL				

Date	Item	Unit	Category	Profit
TOTAL				

DATE	FURTHER NOTE | REMARK

Month :

EGG PRODUCTION

Day	No of chicken	Dark Brown	Light Brown	Blue	White	Green	NOTES
1							
2							
3							
4							
5							
6							
7							
8							
9							
10							
11							
12							
13							
14							
15							
16							
17							
18							
19							
20							
21							
22							
23							
24							
25							
26							
27							
28							
29							
30							
31							
Total							

FEED HISTORY | COMBINATION

Week/Period	Feed Combination	Brand Name	Quantity	Observation

HEALTH RECORD

Day	ID	Injury/Illness	Medication	Dosage	Note

COOP MAINTENANCE

Date	Deep Clean	Light Clean	Notes

MONTH:

EXPENSES

INCOME

Date	Item	Unit	Category	Cost
TOTAL				

Date	Item	Unit	Category	Profit
TOTAL				

DATE	FURTHER NOTE | REMARK

Month :

EGG PRODUCTION

Day	No of chicken	Dark Brown	Light Brown	Blue	White	Green	NOTES
1							
2							
3							
4							
5							
6							
7							
8							
9							
10							
11							
12							
13							
14							
15							
16							
17							
18							
19							
20							
21							
22							
23							
24							
25							
26							
27							
28							
29							
30							
31							
Total							

FEED HISTORY | COMBINATION

Week/Period	Feed Combination	Brand Name	Quantity	Observation

HEALTH RECORD

Day	ID	Injury/Illness	Medication	Dosage	Note

COOP MAINTENANCE

Date	Deep Clean	Light Clean	Notes

MONTH:

EXPENSES

INCOME

Date	Item	Unit	Category	Cost
TOTAL				

Date	Item	Unit	Category	Profit
TOTAL				

DATE	FURTHER NOTE | REMARK

Month :

EGG PRODUCTION

Day	No of chicken	Dark Brown	Light Brow n	Blue	White	Green	NOTES
1							
2							
3							
4							
5							
6							
7							
8							
9							
10							
11							
12							
13							
14							
15							
16							
17							
18							
19							
20							
21							
22							
23							
24							
25							
26							
27							
28							
29							
30							
31							
Total							

FEED HISTORY | COMBINATION

Week/Period	Feed Combination	Brand Name	Quantity	Observation

HEALTH RECORD

Day	ID	Injury/Illness	Medication	Dosage	Note

COOP MAINTENANCE

Date	Deep Clean	Light Clean	Notes

MONTH:

EXPENSES

Date	Item	Unit	Category	Cost
TOTAL				

INCOME

Date	Item	Unit	Category	Profit
TOTAL				

DATE	FURTHER NOTE | REMARK

Month :  **EGG PRODUCTION**

Day	No of chicken	Dark Brown	Light Brow n	Blue	White	Green	NOTES
1							
2							
3							
4							
5							
6							
7							
8							
9							
10							
11							
12							
13							
14							
15							
16							
17							
18							
19							
20							
21							
22							
23							
24							
25							
26							
27							
28							
29							
30							
31							
Total							

FEED HISTORY | COMBINATION

Week/Period	Feed Combination	Brand Name	Quantity	Observation

HEALTH RECORD

Day	ID	Injury/Illness	Medication	Dosage	Note

COOP MAINTENANCE

Date	Deep Clean	Light Clean	Notes

MONTH:

EXPENSES

INCOME

Date	Item	Unit	Category	Cost
TOTAL				

Date	Item	Unit	Category	Profit
TOTAL				

DATE	FURTHER NOTE | REMARK

Month :

EGG PRODUCTION

Day	No of chicken	Dark Brown	Light Brow n	Blue	White	Green	NOTES
1							
2							
3							
4							
5							
6							
7							
8							
9							
10							
11							
12							
13							
14							
15							
16							
17							
18							
19							
20							
21							
22							
23							
24							
25							
26							
27							
28							
29							
30							
31							
Total							

FEED HISTORY | COMBINATION

Week/Period	Feed Combination	Brand Name	Quantity	Observation

HEALTH RECORD

Day	ID	Injury/Illness	Medication	Dosage	Note

COOP MAINTENANCE

Date	Deep Clean	Light Clean	Notes

MONTH:

EXPENSES

Date	Item	Unit	Category	Cost
TOTAL				

INCOME

Date	Item	Unit	Category	Profit
TOTAL				

DATE	FURTHER NOTE | REMARK

Month :

EGG PRODUCTION

Day	No of chicken	Dark Brown	Light Brow n	Blue	White	Green	NOTES
1							
2							
3							
4							
5							
6							
7							
8							
9							
10							
11							
12							
13							
14							
15							
16							
17							
18							
19							
20							
21							
22							
23							
24							
25							
26							
27							
28							
29							
30							
31							
Total							

FEED HISTORY | COMBINATION

Week/Period	Feed Combination	Brand Name	Quantity	Observation

HEALTH RECORD

Day	ID	Injury/Illness	Medication	Dosage	Note

COOP MAINTENANCE

Date	Deep Clean	Light Clean	Notes

MONTH:

EXPENSES

INCOME

Date	Item	Unit	Category	Cost
TOTAL				

Date	Item	Unit	Category	Profit
TOTAL				

DATE	FURTHER NOTE | REMARK

Month :

EGG PRODUCTION

Day	No of chicken	Dark Brown	Light Brown	Blue	White	Green	NOTES
1							
2							
3							
4							
5							
6							
7							
8							
9							
10							
11							
12							
13							
14							
15							
16							
17							
18							
19							
20							
21							
22							
23							
24							
25							
26							
27							
28							
29							
30							
31							
Total							

FEED HISTORY | COMBINATION

Week/Period	Feed Combination	Brand Name	Quantity	Observation

HEALTH RECORD

Day	ID	Injury/Illness	Medication	Dosage	Note

COOP MAINTENANCE

Date	Deep Clean	Light Clean	Notes

MONTH:

EXPENSES

INCOME

Date	Item	Unit	Category	Cost
TOTAL				

Date	Item	Unit	Category	Profit
TOTAL				

DATE	FURTHER NOTE | REMARK

Month :

EGG PRODUCTION

Day	No of chicken	Dark Brown	Light Brown	Blue	White	Green	NOTES
1							
2							
3							
4							
5							
6							
7							
8							
9							
10							
11							
12							
13							
14							
15							
16							
17							
18							
19							
20							
21							
22							
23							
24							
25							
26							
27							
28							
29							
30							
31							
Total							

FEED HISTORY | COMBINATION

Week/Period	Feed Combination	Brand Name	Quantity	Observation

HEALTH RECORD

Day	ID	Injury/Illness	Medication	Dosage	Note

COOP MAINTENANCE

Date	Deep Clean	Light Clean	Notes

YEAR 2

MY FLOCK

Name	ID	Breed	Appearance	Purchase Date	DOB	Lay Date	Cost	Note

MONTH:

EXPENSES

INCOME

Date	Item	Unit	Category	Cost
TOTAL				

Date	Item	Unit	Category	Profit
TOTAL				

DATE	FURTHER NOTE | REMARK

Month :

EGG PRODUCTION

Day	No of chicken	Dark Brown	Light Brown	Blue	White	Green	NOTES
1							
2							
3							
4							
5							
6							
7							
8							
9							
10							
11							
12							
13							
14							
15							
16							
17							
18							
19							
20							
21							
22							
23							
24							
25							
26							
27							
28							
29							
30							
31							
Total							

FEED HISTORY | COMBINATION

Week/Period	Feed Combination	Brand Name	Quantity	Observation

HEALTH RECORD

Day	ID	Injury/Illness	Medication	Dosage	Note

COOP MAINTENANCE

Date	Deep Clean	Light Clean	Notes

DATE	FURTHER NOTE | REMARK

MONTH:

EXPENSES

INCOME

Date	Item	Unit	Category	Cost
TOTAL				

Date	Item	Unit	Category	Profit
TOTAL				

DATE	FURTHER NOTE | REMARK

Month :

EGG PRODUCTION

Day	No of chicken	Dark Brown	Light Brow n	Blue	White	Green	NOTES
1							
2							
3							
4							
5							
6							
7							
8							
9							
10							
11							
12							
13							
14							
15							
16							
17							
18							
19							
20							
21							
22							
23							
24							
25							
26							
27							
28							
29							
30							
31							
Total							

FEED HISTORY | COMBINATION

Week/Period	Feed Combination	Brand Name	Quantity	Observation

HEALTH RECORD

Day	ID	Injury/Illness	Medication	Dosage	Note

COOP MAINTENANCE

Date	Deep Clean	Light Clean	Notes

MONTH:

EXPENSES

INCOME

Date	Item	Unit	Category	Cost
TOTAL				

Date	Item	Unit	Category	Profit
TOTAL				

DATE	FURTHER NOTE | REMARK

Month :

EGG PRODUCTION

Day	No of chicken	Dark Brown	Light Brow n	Blue	White	Green	NOTES
1							
2							
3							
4							
5							
6							
7							
8							
9							
10							
11							
12							
13							
14							
15							
16							
17							
18							
19							
20							
21							
22							
23							
24							
25							
26							
27							
28							
29							
30							
31							
Total							

FEED HISTORY | COMBINATION

Week/Period	Feed Combination	Brand Name	Quantity	Observation

HEALTH RECORD

Day	ID	Injury/Illness	Medication	Dosage	Note

COOP MAINTENANCE

Date	Deep Clean	Light Clean	Notes

MONTH:

EXPENSES

INCOME

Date	Item	Unit	Category	Cost
TOTAL				

Date	Item	Unit	Category	Profit
TOTAL				

DATE	FURTHER NOTE | REMARK

Month :

EGG PRODUCTION

Day	No of chicken	Dark Brown	Light Brow n	Blue	White	Green	NOTES
1							
2							
3							
4							
5							
6							
7							
8							
9							
10							
11							
12							
13							
14							
15							
16							
17							
18							
19							
20							
21							
22							
23							
24							
25							
26							
27							
28							
29							
30							
31							
Total							

FEED HISTORY | COMBINATION

Week/Period	Feed Combination	Brand Name	Quantity	Observation

HEALTH RECORD

Day	ID	Injury/Illness	Medication	Dosage	Note

COOP MAINTENANCE

Date	Deep Clean	Light Clean	Notes

MONTH:

EXPENSES

INCOME

Date	Item	Unit	Category	Cost
TOTAL				

Date	Item	Unit	Category	Profit
TOTAL				

DATE	FURTHER NOTE | REMARK

Month :

EGG PRODUCTION

Day	No of chicken	Dark Brown	Light Brown	Blue	White	Green	NOTES
1							
2							
3							
4							
5							
6							
7							
8							
9							
10							
11							
12							
13							
14							
15							
16							
17							
18							
19							
20							
21							
22							
23							
24							
25							
26							
27							
28							
29							
30							
31							
Total							

FEED HISTORY | COMBINATION

Week/Period	Feed Combination	Brand Name	Quantity	Observation

HEALTH RECORD

Day	ID	Injury/Illness	Medication	Dosage	Note

COOP MAINTENANCE

Date	Deep Clean	Light Clean	Notes

MONTH:

EXPENSES

INCOME

Date	Item	Unit	Category	Cost
TOTAL				

Date	Item	Unit	Category	Profit
TOTAL				

DATE	FURTHER NOTE | REMARK

Month :

EGG PRODUCTION

Day	No of chicken	Dark Brown	Light Brown	Blue	White	Green	NOTES
1							
2							
3							
4							
5							
6							
7							
8							
9							
10							
11							
12							
13							
14							
15							
16							
17							
18							
19							
20							
21							
22							
23							
24							
25							
26							
27							
28							
29							
30							
31							
Total							

FEED HISTORY | COMBINATION

Week/Period	Feed Combination	Brand Name	Quantity	Observation

HEALTH RECORD

Day	ID	Injury/Illness	Medication	Dosage	Note

COOP MAINTENANCE

Date	Deep Clean	Light Clean	Notes

MONTH:

EXPENSES

Date	Item	Unit	Category	Cost
TOTAL				

INCOME

Date	Item	Unit	Category	Profit
TOTAL				

DATE	FURTHER NOTE | REMARK

Month :

EGG PRODUCTION

Day	No of chicken	Dark Brown	Light Brown	Blue	White	Green	NOTES
1							
2							
3							
4							
5							
6							
7							
8							
9							
10							
11							
12							
13							
14							
15							
16							
17							
18							
19							
20							
21							
22							
23							
24							
25							
26							
27							
28							
29							
30							
31							
Total							

FEED HISTORY | COMBINATION

Week/Period	Feed Combination	Brand Name	Quantity	Observation

HEALTH RECORD

Day	ID	Injury/Illness	Medication	Dosage	Note

COOP MAINTENANCE

Date	Deep Clean	Light Clean	Notes

MONTH:

EXPENSES

Date	Item	Unit	Category	Cost
TOTAL				

INCOME

Date	Item	Unit	Category	Profit
TOTAL				

DATE	FURTHER NOTE | REMARK

Month :

EGG PRODUCTION

Day	No of chicken	Dark Brown	Light Brow n	Blue	White	Green	NOTES
1							
2							
3							
4							
5							
6							
7							
8							
9							
10							
11							
12							
13							
14							
15							
16							
17							
18							
19							
20							
21							
22							
23							
24							
25							
26							
27							
28							
29							
30							
31							
Total							

FEED HISTORY | COMBINATION

Week/Period	Feed Combination	Brand Name	Quantity	Observation

HEALTH RECORD

Day	ID	Injury/Illness	Medication	Dosage	Note

COOP MAINTENANCE

Date	Deep Clean	Light Clean	Notes

MONTH:

EXPENSES

INCOME

Date	Item	Unit	Category	Cost
TOTAL				

Date	Item	Unit	Category	Profit
TOTAL				

DATE	FURTHER NOTE | REMARK

Month :

EGG PRODUCTION

Day	No of chicken	Dark Brown	Light Brow n	Blue	White	Green	NOTES
1							
2							
3							
4							
5							
6							
7							
8							
9							
10							
11							
12							
13							
14							
15							
16							
17							
18							
19							
20							
21							
22							
23							
24							
25							
26							
27							
28							
29							
30							
31							
Total							

FEED HISTORY | COMBINATION

Week/Period	Feed Combination	Brand Name	Quantity	Observation

HEALTH RECORD

Day	ID	Injury/Illness	Medication	Dosage	Note

COOP MAINTENANCE

Date	Deep Clean	Light Clean	Notes

MONTH:

EXPENSES

INCOME

Date	Item	Unit	Category	Cost
TOTAL				

Date	Item	Unit	Category	Profit
TOTAL				

DATE	FURTHER NOTE | REMARK

Month :

EGG PRODUCTION

Day	No of chicken	Dark Brown	Light Brown	Blue	White	Green	NOTES
1							
2							
3							
4							
5							
6							
7							
8							
9							
10							
11							
12							
13							
14							
15							
16							
17							
18							
19							
20							
21							
22							
23							
24							
25							
26							
27							
28							
29							
30							
31							
Total							

FEED HISTORY | COMBINATION

Week/Period	Feed Combination	Brand Name	Quantity	Observation

HEALTH RECORD

Day	ID	Injury/Illness	Medication	Dosage	Note

COOP MAINTENANCE

Date	Deep Clean	Light Clean	Notes

MONTH:

EXPENSES

INCOME

Date	Item	Unit	Category	Cost
TOTAL				

Date	Item	Unit	Category	Profit
TOTAL				

DATE	FURTHER NOTE | REMARK

Month :

EGG PRODUCTION

Day	No of chicken	Dark Brown	Light Brow n	Blue	White	Green	NOTES
1							
2							
3							
4							
5							
6							
7							
8							
9							
10							
11							
12							
13							
14							
15							
16							
17							
18							
19							
20							
21							
22							
23							
24							
25							
26							
27							
28							
29							
30							
31							
Total							

FEED HISTORY | COMBINATION

Week/Period	Feed Combination	Brand Name	Quantity	Observation

HEALTH RECORD

Day	ID	Injury/Illness	Medication	Dosage	Note

COOP MAINTENANCE

Date	Deep Clean	Light Clean	Notes

MONTH:

EXPENSES

Date	Item	Unit	Category	Cost
TOTAL				

INCOME

Date	Item	Unit	Category	Profit
TOTAL				

DATE	FURTHER NOTE | REMARK

Month :

EGG PRODUCTION

Day	No of chicken	Dark Brown	Light Brown	Blue	White	Green	NOTES
1							
2							
3							
4							
5							
6							
7							
8							
9							
10							
11							
12							
13							
14							
15							
16							
17							
18							
19							
20							
21							
22							
23							
24							
25							
26							
27							
28							
29							
30							
31							
Total							

FEED HISTORY | COMBINATION

Week/Period	Feed Combination	Brand Name	Quantity	Observation

HEALTH RECORD

Day	ID	Injury/Illness	Medication	Dosage	Note

COOP MAINTENANCE

Date	Deep Clean	Light Clean	Notes

YEAR 3

MY FLOCK

Name	ID	Breed	Appearance	Purchase Date	DOB	Lay Date	Cost	Note

MONTH:

EXPENSES

INCOME

Date	Item	Unit	Category	Cost
TOTAL				

Date	Item	Unit	Category	Profit
TOTAL				

DATE	FURTHER NOTE | REMARK

Month :

EGG PRODUCTION

Day	No of chicken	Dark Brown	Light Brown	Blue	White	Green	NOTES
1							
2							
3							
4							
5							
6							
7							
8							
9							
10							
11							
12							
13							
14							
15							
16							
17							
18							
19							
20							
21							
22							
23							
24							
25							
26							
27							
28							
29							
30							
31							
Total							

FEED HISTORY | COMBINATION

Week/Period	Feed Combination	Brand Name	Quantity	Observation

HEALTH RECORD

Day	ID	Injury/Illness	Medication	Dosage	Note

COOP MAINTENANCE

Date	Deep Clean	Light Clean	Notes

DATE	FURTHER NOTE | REMARK

MONTH:

EXPENSES

INCOME

Date	Item	Unit	Category	Cost
TOTAL				

Date	Item	Unit	Category	Profit
TOTAL				

DATE	FURTHER NOTE | REMARK

Month :

EGG PRODUCTION

Day	No of chicken	Dark Brown	Light Brow n	Blue	White	Green	NOTES
1							
2							
3							
4							
5							
6							
7							
8							
9							
10							
11							
12							
13							
14							
15							
16							
17							
18							
19							
20							
21							
22							
23							
24							
25							
26							
27							
28							
29							
30							
31							
Total							

FEED HISTORY | COMBINATION

Week/Period	Feed Combination	Brand Name	Quantity	Observation

HEALTH RECORD

Day	ID	Injury/Illness	Medication	Dosage	Note

COOP MAINTENANCE

Date	Deep Clean	Light Clean	Notes

MONTH:

EXPENSES

INCOME

Date	Item	Unit	Category	Cost
TOTAL				

Date	Item	Unit	Category	Profit
TOTAL				

DATE	FURTHER NOTE | REMARK

Month :

EGG PRODUCTION

Day	No of chicken	Dark Brown	Light Brow n	Blue	White	Green	NOTES
1							
2							
3							
4							
5							
6							
7							
8							
9							
10							
11							
12							
13							
14							
15							
16							
17							
18							
19							
20							
21							
22							
23							
24							
25							
26							
27							
28							
29							
30							
31							
Total							

FEED HISTORY | COMBINATION

Week/Period	Feed Combination	Brand Name	Quantity	Observation

HEALTH RECORD

Day	ID	Injury/Illness	Medication	Dosage	Note

COOP MAINTENANCE

Date	Deep Clean	Light Clean	Notes

MONTH:

EXPENSES

INCOME

Date	Item	Unit	Category	Cost
TOTAL				

Date	Item	Unit	Category	Profit
TOTAL				

DATE	FURTHER NOTE | REMARK

Month :

EGG PRODUCTION

Day	No of chicken	Dark Brown	Light Brow n	Blue	White	Green	NOTES
1							
2							
3							
4							
5							
6							
7							
8							
9							
10							
11							
12							
13							
14							
15							
16							
17							
18							
19							
20							
21							
22							
23							
24							
25							
26							
27							
28							
29							
30							
31							
Total							

FEED HISTORY | COMBINATION

Week/Period	Feed Combination	Brand Name	Quantity	Observation

HEALTH RECORD

Day	ID	Injury/Illness	Medication	Dosage	Note

COOP MAINTENANCE

Date	Deep Clean	Light Clean	Notes

MONTH:

EXPENSES

INCOME

Date	Item	Unit	Category	Cost
TOTAL				

Date	Item	Unit	Category	Profit
TOTAL				

DATE	FURTHER NOTE | REMARK

Month :

EGG PRODUCTION

Day	No of chicken	Dark Brown	Light Brow n	Blue	White	Green	NOTES
1							
2							
3							
4							
5							
6							
7							
8							
9							
10							
11							
12							
13							
14							
15							
16							
17							
18							
19							
20							
21							
22							
23							
24							
25							
26							
27							
28							
29							
30							
31							
Total							

FEED HISTORY | COMBINATION

Week/Period	Feed Combination	Brand Name	Quantity	Observation

HEALTH RECORD

Day	ID	Injury/Illness	Medication	Dosage	Note

COOP MAINTENANCE

Date	Deep Clean	Light Clean	Notes

MONTH:

EXPENSES

INCOME

Date	Item	Unit	Category	Cost
TOTAL				

Date	Item	Unit	Category	Profit
TOTAL				

DATE	FURTHER NOTE | REMARK

Month :

EGG PRODUCTION

Day	No of chicken	Dark Brown	Light Brow n	Blue	White	Green	NOTES
1							
2							
3							
4							
5							
6							
7							
8							
9							
10							
11							
12							
13							
14							
15							
16							
17							
18							
19							
20							
21							
22							
23							
24							
25							
26							
27							
28							
29							
30							
31							
Total							

FEED HISTORY | COMBINATION

Week/Period	Feed Combination	Brand Name	Quantity	Observation

HEALTH RECORD

Day	ID	Injury/Illness	Medication	Dosage	Note

COOP MAINTENANCE

Date	Deep Clean	Light Clean	Notes

MONTH:

EXPENSES

INCOME

Date	Item	Unit	Category	Cost
TOTAL				

Date	Item	Unit	Category	Profit
TOTAL				

DATE	FURTHER NOTE | REMARK

Month :

EGG PRODUCTION

Day	No of chicken	Dark Brown	Light Brow n	Blue	White	Green	NOTES
1							
2							
3							
4							
5							
6							
7							
8							
9							
10							
11							
12							
13							
14							
15							
16							
17							
18							
19							
20							
21							
22							
23							
24							
25							
26							
27							
28							
29							
30							
31							
Total							

FEED HISTORY | COMBINATION

Week/Period	Feed Combination	Brand Name	Quantity	Observation

HEALTH RECORD

Day	ID	Injury/Illness	Medication	Dosage	Note

COOP MAINTENANCE

Date	Deep Clean	Light Clean	Notes

MONTH:

EXPENSES

INCOME

Date	Item	Unit	Category	Cost
TOTAL				

Date	Item	Unit	Category	Profit
TOTAL				

DATE	FURTHER NOTE | REMARK

Month :

EGG PRODUCTION

Day	No of chicken	Dark Brown	Light Brown	Blue	White	Green	NOTES
1							
2							
3							
4							
5							
6							
7							
8							
9							
10							
11							
12							
13							
14							
15							
16							
17							
18							
19							
20							
21							
22							
23							
24							
25							
26							
27							
28							
29							
30							
31							
Total							

FEED HISTORY | COMBINATION

Week/Period	Feed Combination	Brand Name	Quantity	Observation

HEALTH RECORD

Day	ID	Injury/Illness	Medication	Dosage	Note

COOP MAINTENANCE

Date	Deep Clean	Light Clean	Notes

MONTH:

EXPENSES

INCOME

Date	Item	Unit	Category	Cost
TOTAL				

Date	Item	Unit	Category	Profit
TOTAL				

DATE	FURTHER NOTE | REMARK

Month :

EGG PRODUCTION

Day	No of chicken	Dark Brown	Light Brow n	Blue	White	Green	NOTES
1							
2							
3							
4							
5							
6							
7							
8							
9							
10							
11							
12							
13							
14							
15							
16							
17							
18							
19							
20							
21							
22							
23							
24							
25							
26							
27							
28							
29							
30							
31							
Total							

FEED HISTORY | COMBINATION

Week/Period	Feed Combination	Brand Name	Quantity	Observation

HEALTH RECORD

Day	ID	Injury/Illness	Medication	Dosage	Note

COOP MAINTENANCE

Date	Deep Clean	Light Clean	Notes

MONTH:

EXPENSES

INCOME

Date	Item	Unit	Category	Cost
TOTAL				

Date	Item	Unit	Category	Profit
TOTAL				

DATE	FURTHER NOTE | REMARK

Month :

EGG PRODUCTION

Day	No of chicken	Dark Brown	Light Brown	Blue	White	Green	NOTES
1							
2							
3							
4							
5							
6							
7							
8							
9							
10							
11							
12							
13							
14							
15							
16							
17							
18							
19							
20							
21							
22							
23							
24							
25							
26							
27							
28							
29							
30							
31							
Total							

FEED HISTORY | COMBINATION

Week/Period	Feed Combination	Brand Name	Quantity	Observation

HEALTH RECORD

Day	ID	Injury/Illness	Medication	Dosage	Note

COOP MAINTENANCE

Date	Deep Clean	Light Clean	Notes

MONTH:

EXPENSES

INCOME

Date	Item	Unit	Category	Cost
TOTAL				

Date	Item	Unit	Category	Profit
TOTAL				

DATE	FURTHER NOTE | REMARK

Month :

EGG PRODUCTION

Day	No of chicken	Dark Brown	Light Brown	Blue	White	Green	NOTES
1							
2							
3							
4							
5							
6							
7							
8							
9							
10							
11							
12							
13							
14							
15							
16							
17							
18							
19							
20							
21							
22							
23							
24							
25							
26							
27							
28							
29							
30							
31							
Total							

FEED HISTORY | COMBINATION

Week/Period	Feed Combination	Brand Name	Quantity	Observation

HEALTH RECORD

Day	ID	Injury/Illness	Medication	Dosage	Note

COOP MAINTENANCE

Date	Deep Clean	Light Clean	Notes

MONTH:

EXPENSES

INCOME

Date	Item	Unit	Category	Cost
TOTAL				

Date	Item	Unit	Category	Profit
TOTAL				

DATE	FURTHER NOTE | REMARK

Month :

EGG PRODUCTION

Day	No of chicken	Dark Brown	Light Brown	Blue	White	Green	NOTES
1							
2							
3							
4							
5							
6							
7							
8							
9							
10							
11							
12							
13							
14							
15							
16							
17							
18							
19							
20							
21							
22							
23							
24							
25							
26							
27							
28							
29							
30							
31							
Total							

FEED HISTORY | COMBINATION

Week/Period	Feed Combination	Brand Name	Quantity	Observation

HEALTH RECORD

Day	ID	Injury/Illness	Medication	Dosage	Note

COOP MAINTENANCE

Date	Deep Clean	Light Clean	Notes

YEAR 4

MY FLOCK

Name	ID	Breed	Appearance	Purchase Date	DOB	Lay Date	Cost	Note

MONTH:

EXPENSES

INCOME

Date	Item	Unit	Category	Cost
TOTAL				

Date	Item	Unit	Category	Profit
TOTAL				

DATE	FURTHER NOTE | REMARK

Month :

EGG PRODUCTION

Day	No of chicken	Dark Brown	Light Brow n	Blue	White	Green	NOTES
1							
2							
3							
4							
5							
6							
7							
8							
9							
10							
11							
12							
13							
14							
15							
16							
17							
18							
19							
20							
21							
22							
23							
24							
25							
26							
27							
28							
29							
30							
31							
Total							

FEED HISTORY | COMBINATION

Week/Period	Feed Combination	Brand Name	Quantity	Observation

HEALTH RECORD

Day	ID	Injury/Illness	Medication	Dosage	Note

COOP MAINTENANCE

Date	Deep Clean	Light Clean	Notes

DATE	FURTHER NOTE | REMARK

MONTH:

EXPENSES

INCOME

Date	Item	Unit	Category	Cost
TOTAL				

Date	Item	Unit	Category	Profit
TOTAL				

DATE	FURTHER NOTE | REMARK

Month :

EGG PRODUCTION

Day	No of chicken	Dark Brown	Light Brown	Blue	White	Green	NOTES
1							
2							
3							
4							
5							
6							
7							
8							
9							
10							
11							
12							
13							
14							
15							
16							
17							
18							
19							
20							
21							
22							
23							
24							
25							
26							
27							
28							
29							
30							
31							
Total							

FEED HISTORY | COMBINATION

Week/Period	Feed Combination	Brand Name	Quantity	Observation

HEALTH RECORD

Day	ID	Injury/Illness	Medication	Dosage	Note

COOP MAINTENANCE

Date	Deep Clean	Light Clean	Notes

MONTH:

EXPENSES

INCOME

Date	Item	Unit	Category	Cost
TOTAL				

Date	Item	Unit	Category	Profit
TOTAL				

DATE	FURTHER NOTE | REMARK

Month :

EGG PRODUCTION

Day	No of chicken	Dark Brown	Light Brow n	Blue	White	Green	NOTES
1							
2							
3							
4							
5							
6							
7							
8							
9							
10							
11							
12							
13							
14							
15							
16							
17							
18							
19							
20							
21							
22							
23							
24							
25							
26							
27							
28							
29							
30							
31							
Total							

FEED HISTORY | COMBINATION

Week/Period	Feed Combination	Brand Name	Quantity	Observation

HEALTH RECORD

Day	ID	Injury/Illness	Medication	Dosage	Note

COOP MAINTENANCE

Date	Deep Clean	Light Clean	Notes

MONTH:

EXPENSES

INCOME

Date	Item	Unit	Category	Cost
TOTAL				

Date	Item	Unit	Category	Profit
TOTAL				

DATE	FURTHER NOTE | REMARK

Month :

EGG PRODUCTION

Day	No of chicken	Dark Brown	Light Brown	Blue	White	Green	NOTES
1							
2							
3							
4							
5							
6							
7							
8							
9							
10							
11							
12							
13							
14							
15							
16							
17							
18							
19							
20							
21							
22							
23							
24							
25							
26							
27							
28							
29							
30							
31							
Total							

FEED HISTORY | COMBINATION

Week/Period	Feed Combination	Brand Name	Quantity	Observation

HEALTH RECORD

Day	ID	Injury/Illness	Medication	Dosage	Note

COOP MAINTENANCE

Date	Deep Clean	Light Clean	Notes

MONTH:

EXPENSES

INCOME

EXPENSES

Date	Item	Unit	Category	Cost
TOTAL				

INCOME

Date	Item	Unit	Category	Profit
TOTAL				

DATE	FURTHER NOTE | REMARK

Month :

EGG PRODUCTION

Day	No of chicken	Dark Brown	Light Brow n	Blue	White	Green	NOTES
1							
2							
3							
4							
5							
6							
7							
8							
9							
10							
11							
12							
13							
14							
15							
16							
17							
18							
19							
20							
21							
22							
23							
24							
25							
26							
27							
28							
29							
30							
31							
Total							

FEED HISTORY | COMBINATION

Week/Period	Feed Combination	Brand Name	Quantity	Observation

HEALTH RECORD

Day	ID	Injury/Illness	Medication	Dosage	Note

COOP MAINTENANCE

Date	Deep Clean	Light Clean	Notes

MONTH:

EXPENSES

Date	Item	Unit	Category	Cost
TOTAL				

INCOME

Date	Item	Unit	Category	Profit
TOTAL				

DATE	FURTHER NOTE | REMARK

Month :

EGG PRODUCTION

Day	No of chicken	Dark Brown	Light Brow n	Blue	White	Green	NOTES
1							
2							
3							
4							
5							
6							
7							
8							
9							
10							
11							
12							
13							
14							
15							
16							
17							
18							
19							
20							
21							
22							
23							
24							
25							
26							
27							
28							
29							
30							
31							
Total							

FEED HISTORY | COMBINATION

Week/Period	Feed Combination	Brand Name	Quantity	Observation

HEALTH RECORD

Day	ID	Injury/Illness	Medication	Dosage	Note

COOP MAINTENANCE

Date	Deep Clean	Light Clean	Notes

MONTH:

EXPENSES

Date	Item	Unit	Category	Cost
TOTAL				

INCOME

Date	Item	Unit	Category	Profit
TOTAL				

DATE	FURTHER NOTE | REMARK

Month :

EGG PRODUCTION

Day	No of chicken	Dark Brown	Light Brown	Blue	White	Green	NOTES
1							
2							
3							
4							
5							
6							
7							
8							
9							
10							
11							
12							
13							
14							
15							
16							
17							
18							
19							
20							
21							
22							
23							
24							
25							
26							
27							
28							
29							
30							
31							
Total							

FEED HISTORY | COMBINATION

Week/Period	Feed Combination	Brand Name	Quantity	Observation

HEALTH RECORD

Day	ID	Injury/Illness	Medication	Dosage	Note

COOP MAINTENANCE

Date	Deep Clean	Light Clean	Notes

MONTH:

EXPENSES

INCOME

Date	Item	Unit	Category	Cost
TOTAL				

Date	Item	Unit	Category	Profit
TOTAL				

DATE	FURTHER NOTE | REMARK

Month :

EGG PRODUCTION

Day	No of chicken	Dark Brown	Light Brown	Blue	White	Green	NOTES
1							
2							
3							
4							
5							
6							
7							
8							
9							
10							
11							
12							
13							
14							
15							
16							
17							
18							
19							
20							
21							
22							
23							
24							
25							
26							
27							
28							
29							
30							
31							
Total							

FEED HISTORY | COMBINATION

Week/Period	Feed Combination	Brand Name	Quantity	Observation

HEALTH RECORD

Day	ID	Injury/Illness	Medication	Dosage	Note

COOP MAINTENANCE

Date	Deep Clean	Light Clean	Notes

MONTH:

EXPENSES

Date	Item	Unit	Category	Cost
TOTAL				

INCOME

Date	Item	Unit	Category	Profit
TOTAL				

DATE	FURTHER NOTE | REMARK

Month :

EGG PRODUCTION

Day	No of chicken	Dark Brown	Light Brown	Blue	White	Green	NOTES
1							
2							
3							
4							
5							
6							
7							
8							
9							
10							
11							
12							
13							
14							
15							
16							
17							
18							
19							
20							
21							
22							
23							
24							
25							
26							
27							
28							
29							
30							
31							
Total							

FEED HISTORY | COMBINATION

Week/Period	Feed Combination	Brand Name	Quantity	Observation

HEALTH RECORD

Day	ID	Injury/Illness	Medication	Dosage	Note

COOP MAINTENANCE

Date	Deep Clean	Light Clean	Notes

MONTH:

EXPENSES

INCOME

Date	Item	Unit	Category	Cost
TOTAL				

Date	Item	Unit	Category	Profit
TOTAL				

DATE	FURTHER NOTE | REMARK

Month :

EGG PRODUCTION

Day	No of chicken	Dark Brown	Light Brown	Blue	White	Green	NOTES
1							
2							
3							
4							
5							
6							
7							
8							
9							
10							
11							
12							
13							
14							
15							
16							
17							
18							
19							
20							
21							
22							
23							
24							
25							
26							
27							
28							
29							
30							
31							
Total							

FEED HISTORY | COMBINATION

Week/Period	Feed Combination	Brand Name	Quantity	Observation

HEALTH RECORD

Day	ID	Injury/Illness	Medication	Dosage	Note

COOP MAINTENANCE

Date	Deep Clean	Light Clean	Notes

MONTH:

EXPENSES

INCOME

Date	Item	Unit	Category	Cost
TOTAL				

Date	Item	Unit	Category	Profit
TOTAL				

DATE	FURTHER NOTE | REMARK

Month :

EGG PRODUCTION

Day	No of chicken	Dark Brown	Light Brown	Blue	White	Green	NOTES
1							
2							
3							
4							
5							
6							
7							
8							
9							
10							
11							
12							
13							
14							
15							
16							
17							
18							
19							
20							
21							
22							
23							
24							
25							
26							
27							
28							
29							
30							
31							
Total							

FEED HISTORY | COMBINATION

Week/Period	Feed Combination	Brand Name	Quantity	Observation

HEALTH RECORD

Day	ID	Injury/Illness	Medication	Dosage	Note

COOP MAINTENANCE

Date	Deep Clean	Light Clean	Notes

MONTH:

EXPENSES

INCOME

EXPENSES

Date	Item	Unit	Category	Cost
TOTAL				

INCOME

Date	Item	Unit	Category	Profit
TOTAL				

DATE	FURTHER NOTE | REMARK

Month :

EGG PRODUCTION

Day	No of chicken	Dark Brown	Light Brow n	Blue	White	Green	NOTES
1							
2							
3							
4							
5							
6							
7							
8							
9							
10							
11							
12							
13							
14							
15							
16							
17							
18							
19							
20							
21							
22							
23							
24							
25							
26							
27							
28							
29							
30							
31							
Total							

FEED HISTORY | COMBINATION

Week/Period	Feed Combination	Brand Name	Quantity	Observation

HEALTH RECORD

Day	ID	Injury/Illness	Medication	Dosage	Note

COOP MAINTENANCE

Date	Deep Clean	Light Clean	Notes

SUPPLIER CONTACT DETAILS

Name	
Business	
Phone	
Address	
Note	

Name	
Business	
Phone	
Address	
Note	

Name	
Business	
Phone	
Address	
Note	

Name	
Business	
Phone	
Address	
Note	

Name	
Business	
Phone	
Address	
Note	

Name	
Business	
Phone	
Address	
Note	

Name	
Business	
Phone	
Address	
Note	

Name	
Business	
Phone	
Address	
Note	

Name	
Business	
Phone	
Address	
Note	

Name	
Business	
Phone	
Address	
Note	

Made in the USA
Columbia, SC
03 December 2024

48339345R00087